Continents and Climates

Contents

Puffin Books

The Round Earth

The world we live on is like a huge spaceship spinning around the sun. Just like a spaceship, it carries everything needed to keep its passengers alive. It has air to breathe, water to drink and food to eat. All living things on earth are its passengers; plants, animals and people, and all of them need a place to live. The earth seems very large, but there are vast areas of it covered by oceans and places where no one can live and nothing will grow. The land where life is possible makes up a surprisingly small part of our world.

The earth was formed about 4,500,000,000 years ago. Most scientists think it was made from a giant cloud of hot gases swirling in space. The gases spun slower and slower until they became a hot liquid and came together in a ball. A hard crust formed on the surface. This crust covers all the earth; it is the land we live on and even makes up the land which lies under the sea. Deep inside the earth however, the liquid is still hot. It is made up of melted rock which seethes inside the crust and sometimes makes *earthquakes* and *volcanoes* on the surface.

Only one quarter of the earth's crust lies above sea level. It is divided into huge pieces called *continents* which are surrounded by the oceans of the world. The continents are made up of different kinds of land, high and low, hot and cold, where plants and animals live. The oceans also contain plant and animal life of their own, but all life, wherever it is, depends on the heat and light of the sun.

The earth spins as it circles around the sun. This makes our day and night. It is day when our side of the earth faces the sun and night when it turns away. If the earth did not spin, one side would

Even in very wet climates, man has learned how to build houses which keep him dry.

It is daytime for the side of the earth facing the sun and night-time on the other side.

always face the sun and burn up in its hot rays while the other side became cold and lifeless without any heat or light at all. The earth spins so slowly that it takes 24 hours to make one complete turn.

In summer, the days get longer and the weather gets warmer. In winter, the days become shorter and the nights longer, the land does not get as much warmth from the sun and the weather grows colder. These changes of temperature throughout the year bring different weather in each season, and this makes up the *climate* of the region. The climate decides what kind of plants will grow, what sort of clothes people wear and what type of houses they build.

Over long periods of time, many animals and plants have been forced to adapt to life in regions with severe climates or poor land. Man alone can choose to live where he wants. He can cross the oceans and seas to other lands. He can clear the jungles and make the deserts green to create a new home for himself. From the fishers and hunters of the Arctic to the herdsmen who roam the deserts, he has shown that life is possible in all but the worst climates of our spaceship earth.

Continents

The oceans of the world cover nearly three quarters of the earth's surface. They join together as one body of water which stretches all over the world. The remaining quarter of the earth's surface rises up above the oceans to make dry land. It varies in size from huge continents to tiny islands but these pieces of land have something in common. They are all surrounded by water.

The six continents are: Europe and Asia, Africa, North America, South America, Australia and Antarctica. There are also many islands. Greenland, New Zealand, Great Britain and Japan are some of the island countries of the world, but there are thousands of others scattered around the globe.

Except for Antarctica, which is almost completely covered by a thick sheet of ice, the continents are made up of many different types of land. It can range from high mountains to broad flat lowlands, and can be covered by grasslands and forest, deep thick jungle, hot desert sands or polar ice. Each of these environments has its own special plant and animal life.

The surface of the land can be covered in many different ways. It can be bare like the rough rocky sides of the mountains, or buried beneath the ice which spreads over the polar regions. The deserts of the world are sometimes covered by nothing but sand and stones which stretch for many kilometres across the dry desolate land, while the swamps are often so wet that their ground is too soft to walk on. All these regions have some forms of life, but the rich soil which covers the farming lands is the best growing land in the world.

Soil is the thin layer of loose earth which covers the bed rock of the land. It can be many metres deep in river basins and as thin as a few centimetres on mountain slopes and hillsides. Soil contains all things that are needed for plants to grow. It stores heat from the sun and holds minerals and water which plants use for food. It is made of tiny broken pieces of rocks and sand and is mixed with bits of decayed plants and animal remains called *humus.*

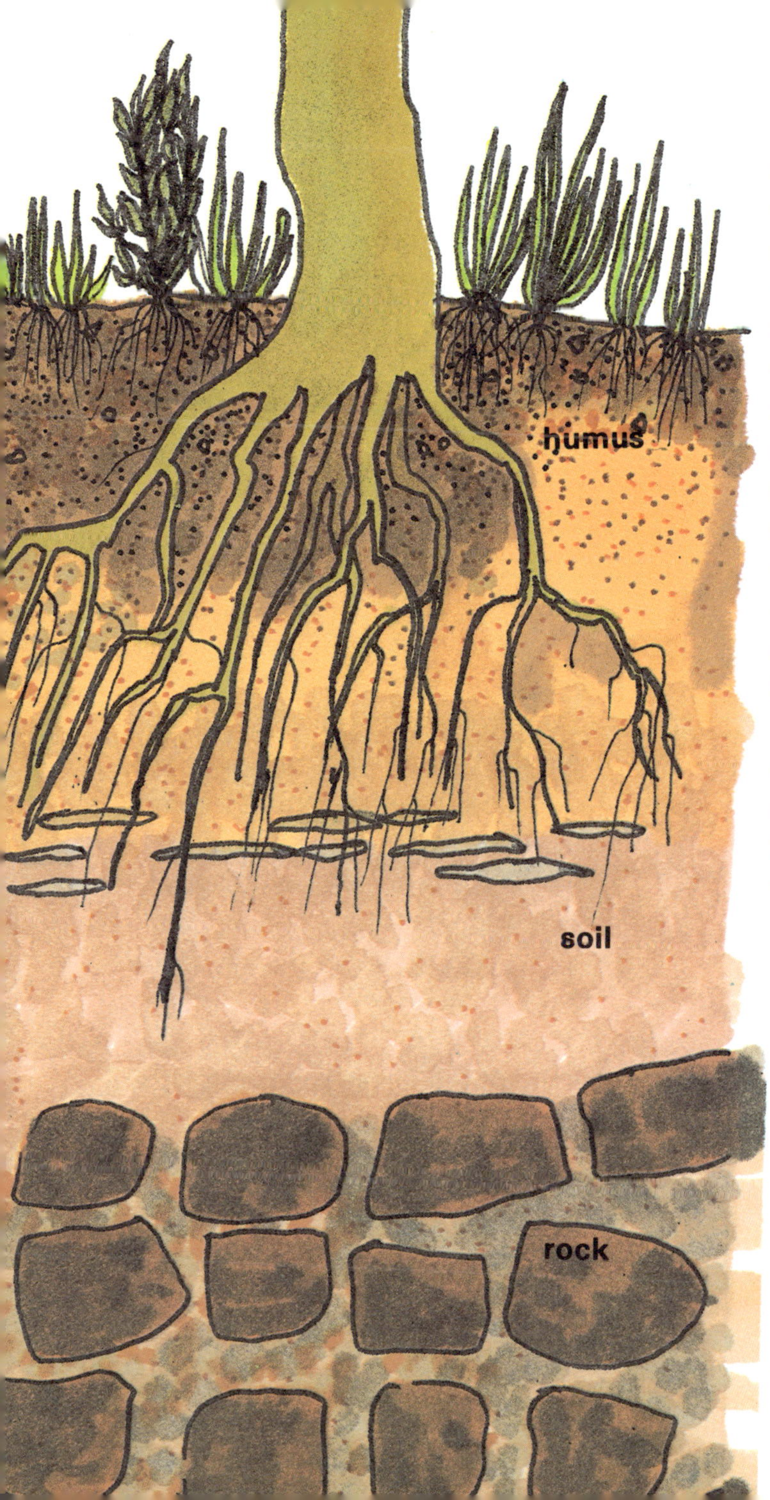

Water is the most important ingredient in the soil. Plants can not live without water. This can be seen in times of drought when no rain falls and the soil becomes dry. Then lush green fields wither and turn brown as the plants are slowly starved of food. Because water is so important, man has learned to use it to make land rich. It is possible to turn parts of dry deserts into green farm land by diverting river water on to them. This is called *irrigation* and has been used in countries like Israel and Australia to create rich land where nothing could grow before.

Man can not usually decide how much water comes to the land because rainfall is controlled by the weather. The kind of weather that a region has during the year makes up its *climate.*

Tree roots growing through layers of soil.

Climates

All over the earth, different places have different kinds of weather. Some regions have hot weather and others have cold, sometimes the weather is dry and sometimes it is wet. There are places on earth where the weather remains the same all year round. In other regions, the weather changes daily and the temperature ranges from hot summers to cold winters. These differences affect everything, the kind of clothes we wear, the food we eat and the houses we live in. They make up the weather zones called climates.

Desert climates are hot with very little rainfall. Polar climates are very cold most of the year. The rain forests of Africa, South-East Asia and South America have warm weather all the year round and so much rain that the plants grow into a thick tangled jungle. The weather in temperate climates changes often through the year, but it never becomes too hot or too cold for plants and animals to live.

Life on earth must have sunlight and water. Plants can not live without moisture for their roots and warmth and light to make them grow. Most animals need plants to eat and water to drink. The amount of sunlight and rainfall that is found in the different climates of the world decides what kinds of plants and animals can live there. In places where there is not enough sun or rain, it is hard for anything to survive.

Some of the soils in desert regions are rich in minerals and the sun shines bright, but crops can not be grown there because rainfall is rare. The Antarctic continent is covered with water but it is frozen in a layer hundreds of metres thick and the sunlight is so pale that the ice never melts and the land remains cold and bare. In tropical regions however, the plants grow quickly to great sizes because they have all the sun and water they need all the year round.

Climate zones circle the earth in belts. The zones near the equator are warm because that is where the sun's rays are strongest. The regions at the poles are cold because the sunlight is weak. The shape of the land and the winds that blow across it also affect climates. Mountain tops, even at the equator, are very cold and lands which are swept by dry winds can never be anything but deserts.

There are many climates between the bitter cold of the poles and the baking heat of the deserts. This has given us an amazing variety of life on earth. In each climate, plants, animals and people have adapted to the conditions so that they can live. Man however, is the only living thing that can control his environment.

Inventions like central heating and air conditioning can make life possible for him anywhere on earth. People must learn to care for wildlife too, for each special species has adapted over millions of years to life in its climate, and man's interference can thoughtlessly destroy the life of a whole region in the space of a few years.

Climates of the World

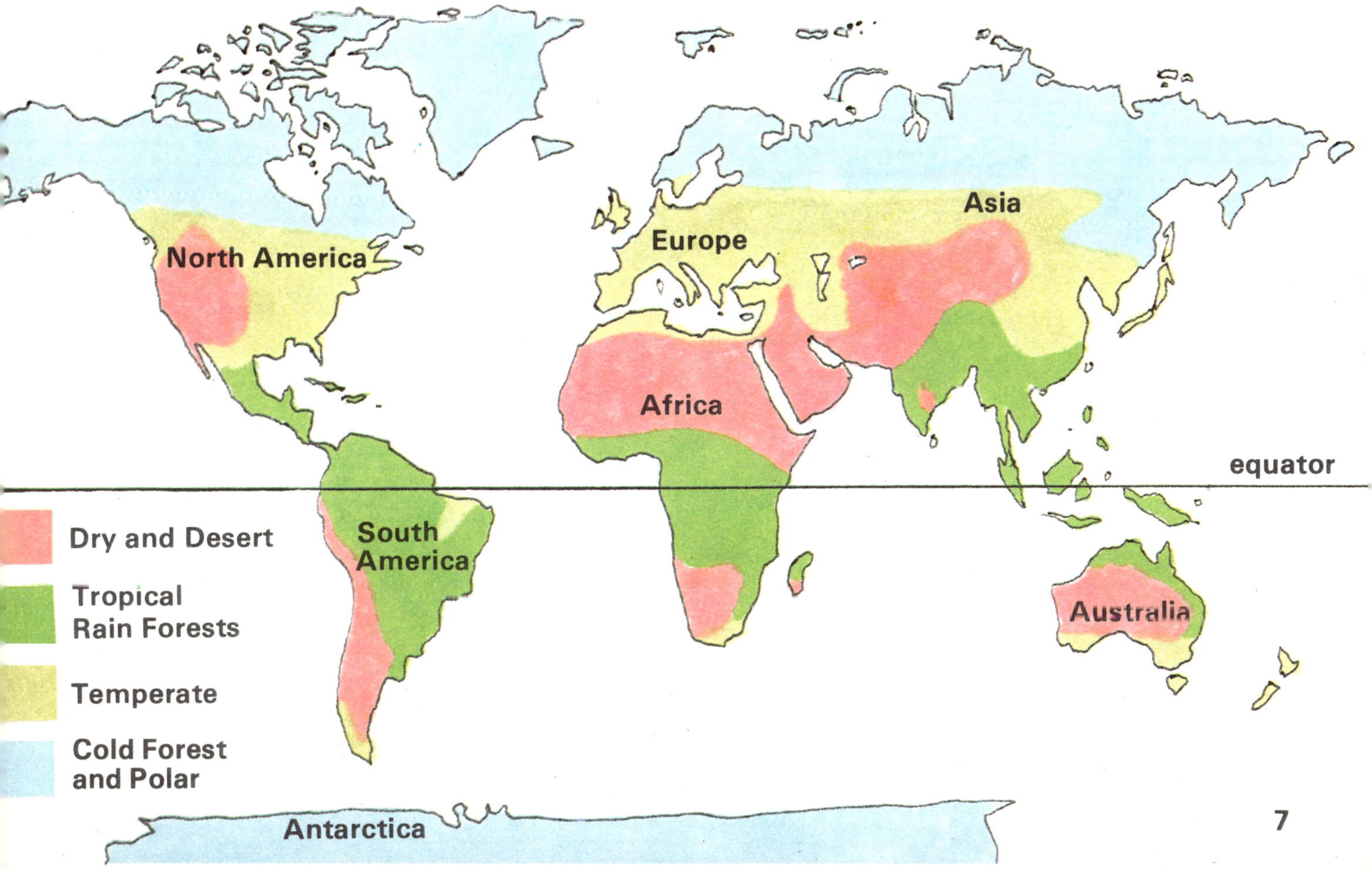

Tropical Rain Forests

The dense rain forests of the world are in warm wet lands near the equator. Parts of Africa, South America, South-East Asia, India and Australia are covered by these thick jungles. The plant and animal life may vary from place to place but they all have a similar climate.

The temperature in rain forest climates is always warm, there is no winter, and the only time it gets cooler is at night. The sun is strong and bright and the yearly rainfall is so great that plants grow quickly to amazing sizes.

Trees grow high and close together. Their spreading branches block out the light from the ground below. Plants that grow closer to the ground, tangle around each other and twine around the trees as

they fight up toward the sunlight. This undergrowth can become so thick that often the only way through the jungle is made by the rivers. They are kept broad and full by the heavy rains which sometimes come every day.

The rich plant life in the jungles makes good food for animals, but the undergrowth is usually so thick that larger animals must live in the trees. In Africa, monkeys and chimpanzees live in large groups in the upper branches. In South-East Asia, there are orang-utans and gibbons. Rain forests also have many

snakes, brightly coloured birds like parrots, toucans and macaws and large clouds of insects. The dense tangle of leaves holds in moisture so that the ground is damp.

The rich land of the rain forest regions could be made more useful to man if food was grown on it but the jungles must be cut down carefully. If the land is stripped too quickly, the pouring rains will soon wash away the rich soil and the countryside could become a desert instead.

In Thailand, valuable trees are cleared from the jungles.

Our Amazing World . . .

Most animals live on plants, but there are some plants that live on animals. The sundew, pitcher plant and venus fly-trap have ways of catching insects which land on them. Some have sticky stems and others have slippery inner chambers or leaves that close up to trap the insects inside. These plants digest the insects to get the minerals which they can not get from the soil because the tropical rains have washed the minerals away.

Deserts

One fifth of the land in the world is covered by deserts and most are close to the equator. Deserts have so little rainfall that they have no trees and very few plants. Bare earth, stones or hot sands cover the surface. Strong winds sweep across the land during the day raising the dry soil into dust storms and sometimes forming curious and changing shapes.

Clouds do not often appear in desert skies because the winds are dry. There is no protection from the burning sun which beats down upon the land, making it unbearably hot during the day. At night there are no clouds to trap the heat in and the land quickly becomes very cold. Not all deserts are hot all year. The Gobi Desert in Mongolia and the Patagonian Desert in Argentina have cold winters, but they are still too dry for most plants to grow.

Some animals have adapted to life in the harsh climate. Desert rats and foxes dig deep burrows in the ground where they can shelter from the midday heat, sometimes only coming out at night. Some animals do not need to drink water and live on the moisture they can get out of the sparse plant life. Plants have either adapted to the long dry spells or learned to store water inside their stems.

The bushmen of the Kalahari Desert in Africa and the aborigines of Australia gather roots and berries and hunt animals. Other desert people called *nomads* live in Central Asia and Africa. They tend flocks of goats which they move from place to place to find enough food.

A desert oasis can sometimes disappear if its water supply dries up.

In some parts of the desert, life is only possible in a settlement called an *oasis.* They are lush green places like cool islands in the sea of burning sand. In an oasis, every scrap of ground is used to grow food such as dates, figs, olives and apricots, and palm trees are planted to shade the houses from the sun.

Rainfall is so rare in the desert that plants grow slowly. They can easily be killed off by nomads who graze too many animals or by accidental grass fires. Without plants, the topsoil at the edge of the desert blows away in the wind, leaving the bare ground behind, and the deserts can grow larger. In this way, man has carelessly created large parts of the Sahara desert and vast areas of semi-desert in other regions of the world.

Most deserts are too hot and dry for plants to grow. There are other regions on the earth where it is too cold for plants to survive. These are the *Polar Regions.*

Our Amazing World . . .

Some deserts on earth were not always dusty and dry. Thousands of years ago, the Sahara was once a green country with a large population. Cave paintings and fish bones found in the centre of the desert show us that it was a fertile land with large flowing rivers and many animals. The weather changed about 5,500 years ago and the heavy rains stopped. The land began to dry out and the topsoil blew away leaving only the sand and rocks behind. Today the desert is still getting drier. Wells have to be dug deeper to find any water under the hot dusty ground.

Polar Regions

The lands and oceans surrounding the north and south poles are the coldest climates on earth. Even in milder parts, the ground is buried beneath snow for most of the year and the summer sunlight is watery and weak. Wildlife is very scarce, and there are huge areas where life is not possible at all.

The polar regions are cold because they lie on parts of the earth that are curved away from the sun. The sun's strong rays do not shine straight down on the surface like they do at the equator. The sunlight is weak because it has been filtered out by the curving atmosphere above. The sun never rises very high in the sky.

The angle of the sun also makes a great difference to the seasons. In winter, as the days get shorter, the land becomes bitterly cold. The sun appears lower and lower in the sky. At the poles, it does not rise at all in mid winter. Summertime only lasts for a few months but the days are very long. In mid summer, the sun never sets and the polar regions are sometimes called "The land of the midnight sun".

The north polar region covers millions of square kilometres of the Arctic Ocean. The sea is frozen into a giant sheet of ice for most of the year, and only animals like polar bears and seals can survive in this icy wilderness. They eat fish which live in great numbers in the cold waters. During the summer, the ice breaks up in places and huge pieces called *icebergs* float out to sea.

Further south, Greenland and the northern parts of Canada, Europe and Russia also have a polar climate. For most of the year, the ground is frozen hard and covered with a thick blanket of snow. The summer season only lasts from June, when the soil thaws, until August when it becomes frozen again. Arctic plants must complete a whole growing cycle in this short time. It is too cold for trees to grow at all.

People who live on these cold plains either hunt and fish like the Eskimos of Canada or tend reindeer like the Laplanders of Northern Europe. They do not live settled lives because they must follow the animals who have to move on frequently to find enough to eat.

In some places valuable minerals have been found beneath the land. Huge oil deposits have been discovered in Alaska and gold and uranium in Russia.

Pipelines, roads and railways are being built into these areas to bring out the minerals. Naturalists hope that the animals and plants will not be destroyed by this invasion of men and machines.

The South Polar Region covers the huge continent of Antarctica. Here the ice and snow are so deep that they never melt. In the centre of the continent the ice sheet is one thousand metres thick. There are only a few small areas around the coast where plants can grow, and most of the animals live on fish. Petrels, sheathbills and penguins live in large colonies on the ice and rocks around the coastline. The only people on Antarctica are the scientific research teams who live in specially heated buildings buried deep in the snow. They hope to learn more about the climate and conditions of these coldest regions on earth.

Our Amazing World . . .

The polar regions can become so cold that tears freeze on people's cheeks and water poured from a glass turns into lumps of ice before it reaches the ground. Quick-frozen food was discovered in these conditions by Clarence Birdseye who was fond of fishing in Alaska. He found that the fish he caught froze solid as soon as he pulled them out of the water. Even after keeping them in a freezer for a long time, his fish were still fresh when they were thawed out. He soon tried quick-freezing vegetables and other foods and started the frozen food industry.

Temperate Zones

Throughout the year, the climates at the equator and the poles stay very much the same. There are regions in between the two however, where the climate changes a great deal. These are the temperate zones which cover broad areas of the earth. The weather in these regions changes quickly but they produce most of the world's food.

Many different kinds of weather are found in temperate zones because they are places where air currents from hot and cold countries mix. Cold air from the polar regions can meet with warm air currents from the tropics to make fierce storms which rage across the countryside. The weather can change very quickly too so that Arctic winds can bring snows even in the summertime and dry air can bring droughts. These rapid changes make temperate climates very unpredictable.

As well as the great variety of weather, some temperate zones also have very extreme seasons. In the centre of the continents where the temperatures are not protected by the oceans, the summer heat can be as intense as semi-desert and the winters as cold as the Arctic.

People who live in these regions have learned to live with the changes. Their houses are designed to be warmly heated in winter but cool in the summer. They have different clothes for different seasons too. Heavy waterproof coats to keep out the wintery cold and rain, and lightweight clothes for the hot days of summer, are part of everyone's wardrobe. Food also changes with the seasons, especially fruits and vegetables which are sometimes only available for a few weeks each year. The changing seasons make life very different

from other climates of the world where people eat the same foods and wear the same kind of clothes all the year round.

The long warm summers and regular rainfall make temperate zones the best land for farming in the world. Huge forests, wide plains of wheat and maize, fields of cotton, potatoes and tobacco,

Planting in the Spring

large orchards of fruit trees and grazing land for cattle and sheep are some of the types of farming land which cover the countryside. The seasonal changes mean that most crops are planted in the spring. They grow steadily through the summer and are ready for harvest in the autumn. Ploughing then begins so that the fields are ready for new crops again in the spring. Some of the plants that live through the year also change with the

seasons. Shade trees like ash, maple, oak and sycamore drop their leaves in autumn and grow new buds in spring. These same trees would stay green all the year round in the tropical rain forests but they have adapted to life in the changeable temperate zones.

Our Amazing World . . .

Farmers in the temperate zones have to be careful about which crops they plant. They sometimes change crops each year so that the soil stays fresh or use fertilizers to put minerals back into the ground. In the early 1900's the farmers in the mid-western United States did not know how to care for their land. They planted the same crops too often in their fields. The land became so dry that the rich topsoil blew away and vast areas of this rich region turned into a dusty desert.

Harvesting in the Autumn

Things to Do

A Sundial

The earliest clocks invented by man were sundials which use shadows to show the time. It is easy to make one which you can use for your own timekeeper. Find an open spot in your garden or playground, and push a stick into the ground. When the sun shines, the stick will cast a shadow on the ground. Use a clock or watch to tell the time. Every hour on the hour, paint or chalk a mark on the ground along the shadow line. Write the hour next to the mark. If you miss an hour or two, you can always come back and do them the next day. When all the marks are in place, your sundial will tell you the time all day.

Eskimo Sunglasses

Sunlight reflecting off the white Arctic landscape can be very dazzling. Eskimos invented "sun-glasses" hundreds of years ago to protect their eyes. Here is how you can make some yourself. Take a strip of paper about 5 x 60 cm.

1. Fold the strip in half.
2. Cut out a shape for your nose.
3. Fold it again about 3 cm from the end.
4. Use a pair of scissors to make two cuts about 1 mm apart.
5. Unfold the strip and pull out the inside of the eye slits.
6. Wrap the strip around your head and tape or glue it at the back.

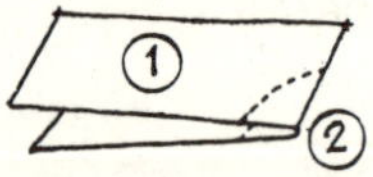

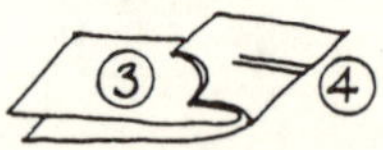

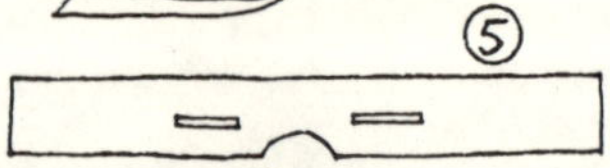

Changing Clothes and Seasons

Many animals have adapted their bodies to a wide range of temperatures. People can only change their clothes. Make a list of the clothes you would wear for a summer swim. Make another list of the clothes you would need for playing in the snow. Which list is longer? If it is possible to collect the two sets together, you could weigh them to see which is heavier. Find out more about animals which are useful to man because of their ability to live in extreme climates, like the camel, husky dog, reindeer, elephant etc. You could make a chart with columns like these to record your information:

Animal	Climate	Special Abilities

A Jungle Jar

You can make your own tropical rain forest in a large jam jar. First find an old jar that has been washed out. Collect a handful of stones and put them in it. Cover the stones with about 3cm of earth from the garden. Collect some seeds like grass, mustard and cress, assorted flowers or even weeds like dandelion, shepherd's purse and soldier's buttons. A greater variety will make your jungle more exotic. Sprinkle the seeds on the earth in the jar and use a pencil to stir them in or scatter some more earth over the top to cover them up. Pour enough water on the earth to make it damp but not too wet. Now use a hammer and a large nail to make some holes in the lid of the jam jar. (Lay the lid on an old scrap of wood before you bang the holes in it.) Put your jungle jar on a warm sunny window sill, and see what happens after a few days.